MW00464875

WITHDRAWN FROM COLLECTION
OF SACRAMENTO PUBLIC LIBRARY

OCT 8 1987

INVENTIONS THAT CHANGED OUR LIVES

Rubber

Shaaron Cosner

Walker and Company
New York

Other titles in the Series

Copyright © 1986 by Shaaron Cosner

All rights reserved. No part of this book may be reproduced or transmitted in any form or by any means, electric or mechanical, including photocopying, recording, or by any information storage and retrieval system, without permission in writing from the Publisher.

First published in the United States of America in 1986 by the Walker Publishing Company, Inc.

Published simultaneously in Canada by John Wiley & Sons Canada, Limited, Rexdale, Ontario.

Library of Congress Cataloging-in-Publication Data

Cosner, Shaaron.
  Rubber.

  (Inventions that changed our lives)
  Includes index.
  Summary: Examines the nature and many applications of one of the world's most versatile materials, describing its discovery and the various experiments and developments increasing its usefulness.
  1. Rubber—Juvenile literature. [1. Rubber]
I. Title. II. Series.
TS1890.C76 1986   678′.2   86-5603
ISBN 0-8027-6653-6
ISBN 0-8027-6654-4 (lib. bdg.)

Printed in the United States of America

10 9 8 7 6 5 4 3 2 1

Contents

Rubber products are used extensively on ships. Here an aircraft carrier is about to take on fuel from an approaching tanker. The rubber hoses are in readiness. *(Photo courtesy U. S. Navy)*

Introduction

IT IS SAID that if all the rubber in the world were destroyed, civilization would be set back 150 years! Without rubber tires, for example, many businesses would cease to function. Cars and trucks would rust away in garages. Manufactured goods would rot in warehouses. Roads would be empty. Gasoline stations would be abandoned, and drive-in windows at banks and fast-food restaurants would close. People would take few vacations or trips and would have to rely almost totally on neighborhood stores for groceries, clothing, and entertainment.

Without rubber there would be no more games such as football, baseball, basketball, golf, tennis, squash, or handball unless athletes went back to using inflated pigs' bladders for balls. What would monster movies be like without the rubber models used as props? Old faucets would drip without rubber washers and underwear would droop without elastic.

Wars would be lost without rubber. Field guns, tanks, armored cars, planes, field kitchens, and hospitals roll on rubber tires. Soldiers march on rubber heels. In one war alone, World War II, each military vehicle averaged three hundred rubber parts; bombers used four hundred. Each battleship needed thirty-five thousand tons of rubber. Without rubber, the Allied nations would have been desperate. When Japan captured all the main sources of natural rubber in Asia, the American and Allied nations discovered that without rubber their lives would be changed forever. Fortunately, before that could happen, synthetic rubber was invented and the Allies won the war. But rubber became more necessary than ever and, today, people use millions of tons of rubber each year.

What makes this material so popular? One reason is that it can be molded into virtually any shape or size imaginable. This means it can be made into anything from huge dock fenders for ships to small caps for medicine bottles. It can be whipped into a foam as soft as whipped cream, but firmer. It can be used as a coating for products such as gloves or dolls. It can be rolled into sheets or film, then cut into fine tapes or threads.

Rubber can be changed a thousand ways by mixing it with other materials. It can be woven with fibers such as nylon or polyester to make strong, firm, yet elastic materials for things like athletic uniforms and swimming suits. It can be bonded to metals and become as strong as steel.

Rubber is one of the most versatile materials in the world. It can be stretched to ten times its original size. It can be squeezed, twisted, and knotted, and it always springs back to its original shape. It is waterproof and airtight. It can also be sterilized, making it perfect for medical supplies.

Plastics have some of these same qualities, but plastic products have never been able to match the elasticity or durability of rubber, especially for outdoor use in cold weather when plastic becomes rigid and cracks. And some plastic products melt under extreme heat. Certain synthetic rubbers are better than natural rubber for the uses for which they have been designed, but although each synthetic rubber is "rubberlike," no single one has all the qualities of natural rubber.

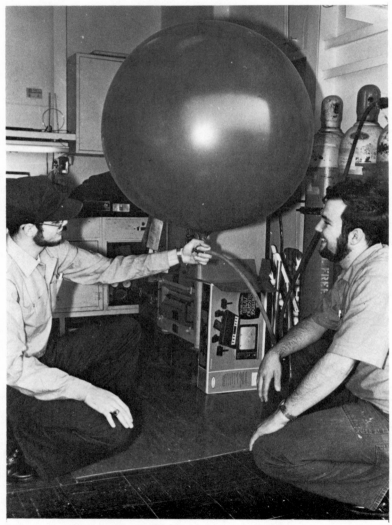

This rubber weather balloon, only partially inflated, is used by the United States Navy. *(Photo courtesy U. S. Navy)*

1

Miracle Gum

OVER FIVE HUNDRED years ago, Indians in
Central and South America played a game that
was a combination of basketball, football, and jai
alai. The object of this game, called Tlachtli (Tlak-
tle), was to knock a ball through a stone ring
without the players using their hands. They hit
the ball with their knees, elbows, shoulders, or
other parts of their bodies. Early explorers who
watched this game discovered that the ball was
made from a white, thick liquid that flowed from a
certain type of tree that grew in the area.

Reports of this wonderful substance began to
trickle back to Europe. A Spanish historian re-
ported that Columbus and his men had seen a ball
"made of the Gum of a Tree which though heavy
would fly and bound better than those fill'd with
Wind in Spain." Cortez's secretary described how
the balls were made. He said the gum of these

special trees was hardened by heat and then molded into balls that would then bound "incredibly into the ayer."

This material would eventually be called rubber because it could be used to *rub out* mistakes. It had actually been used by the Indians as early as A.D. 1050. By the time Cortez and Columbus discovered its existence, the Indians were spreading it on clothes to make them waterproof. They knew how to hollow out balls and fill them with water to make primitive squirt guns. They offered rubber to the gods because they liked the huge black clouds of smoke that resulted when rubber was burned.

Nothing much came of the discovery of rubber until 1735. Then a Frenchman, Charles Marie de la Condamine, while on an expedition to Peru, sent a sample of this strange "elastic gum" to the Academy of Science in Paris. He told the scientists that the Indians called it caoutchouc from *caa* meaning wood and *o-chu* meaning to flow or weep.

Meanwhile Condamine's friend, François Fresneau, had found rubber trees growing in French Guinea along the coast of South America. He was very excited about the new material. He made himself a pair of shoes and a waterproof overcoat from the rubber. He believed that one day it would be used to make divers' suits, water bottles, tarpaulins, pump hoses, and bags for travelers or soldiers who needed to carry their own food.

Reports of the new material were viewed as interesting, but there was a major problem: there was no way to ship rubber in its original liquid state. Only dry rubber could be shipped. As soon as samples of hard rubber arrived in Europe, chemists began searching for a substance that would make it liquid again. They found that turpentine and ether would do this.

At first rubber was a novelty. The king of Portugal sent his boots to Brazil to be waterproofed. The Indians there had learned to make shoes by pouring warm rubber over their feet (used like molds), peeling it off, then hanging it to dry in the sun. People in Europe tried to copy the process. Others used rubber to rubberize silk for some of the new hydrogen balloons being flown. It was also used to waterproof leather, cotton, linen, and wool.

People were eager to invest money in the new "miracle" material. Factories opened to produce a variety of rubber products. A rubber factory in Paris made elastic bands for garters. In England a coach maker named Thomas Hancock opened a rubber factory to manufacture waterproof clothes for coach passengers because travel on the unpaved roads was such a dirty experience. He also produced rubber wrists on gloves to keep them from slipping off and rubber for the tops of pockets to keep pickpockets out.

About this same time, a Boston man, Edwin M. Chaffee, invented three major machines for producing rubber products. One he called the "pickle" which shredded the leftover rubber so it could be used again. (Later this was called a masticator.) Another was a mill that used rollers heated by steam to force dry rubber sheets through. The third was a calender that pressed the rubber between revolving cylinders to make it smooth. Chaffee helped open the first rubber company in the United States and used his machines to make shoes, life preservers, and wagon covers.

Despite the hundreds of products being made from rubber, customers soon learned that it was not perfect. Even at room temperature it was sticky. In hot weather it became stickier and softer. In cold weather it became hard and stiff.

People did not like walking around in winter in shoes as hard as iron boots. They were embarrassed in the summer when their garters and suspenders suddenly sagged or their shoes stuck to the pavement. They found that clothes made of rubber made strange noises when they walked. Raincoats stood up by themselves.

Companies began failing. People were suddenly out of work. Factories were abandoned. Manufacturers tried desperately to invent a new process that would keep rubber soft and pliant regardless of the temperature. But it would be up to one man, Charles Goodyear, to find the solution.

2

Charles Goodyear

IN 1819 CHARLES MACINTOSH, a Scotsman, found he could sandwich two of the sticky layers of rubber between layers of cotton to make a better product. He used his new process to make raincoats which came to be known as mackintoshes.

About this same time, the chemist Michael Faraday figured out the chemical makeup of rubber. He said rubber was made of five carbon atoms and eight hydrogen atoms. His discovery helped pave the way for solving the mysterious qualities of rubber.

Meanwhile Charles Goodyear had become fascinated with rubber. Charles came from a family of inventors. His father had been involved in a patent for button making and had invented Goodyear's Patented Spring Steel Hay and Manure Fork. Charles and his father used their knowledge of practical items to open the first hardware store in the United States and were quite well-to-do until it failed.

Charles Goodyear had first become familiar with rubber when he was a young boy in elementary school. After he had grown up, he happened to go into the salesroom of the Roxbury India Rubber Company in New York where he saw a display of rubber life preservers. Later he read in the newspaper that twenty people drowned every hour throughout the world. According to one of Goodyear's biographers, Ralph F. Wolf, Goodyear couldn't sleep that night. When his wife, Clarissa, asked what was wrong, he replied, "How can I sleep, when so many of my fellow creatures are passing into eternity every day and I feel that I am the man that can prevent it?"

The owner of the rubber company had told Goodyear how difficult it was to deal with this crazy product. He had had thirty thousand dollars' worth of shoes returned—all stuck together and smelly—because the rubber had become too warm. The shoes had to be buried in a pit. The owner could tell that Goodyear was very interested, but he told him there was no use working on the products when the material they were made of was so bad.

"Go home and find a way to make better rubber," he told Goodyear.

That proved to be the most important conversation in Goodyear's life. With no experience in chemistry, he set up a laboratory in the kitchen of his small house in Philadelphia. His only tools were his wife's kitchen utensils. He made a mess of the kitchen every day mixing with rubber anything he could find—salt, pepper, sugar, sand, castor oil, soup, cream cheese, black ink, and other common household items. He baked the results of his experiments in the oven. He figured he would have the problem solved in a month or two.

Months went by and still Goodyear had not solved the rubber problem. He spent every cent he had on his experiments and was often put in jail for bad debts. He was always an optimist, however, and when he was able to produce an improved sheet of rubber in 1834, he felt ready to manufacture rubber goods. That winter, with the help of Clarissa and their daughters, Ellen and Cynthia, he made several hundred pairs of shoes. They were stored away for a while to test their quality. When the warm weather returned, the shoes turned into a mass of melted gum. Goodyear's friends refused to help him any further. Merchants shut off his credit. He sold his furniture to pay bills, and his wife spun linen to post as security for the rent on their cottage while Goodyear himself took off for New York to continue experimenting.

There friends supplied him with a room, and a druggist gave him the necessary supplies to continue. His family joined him eventually in a small house on Staten Island. Charles made trips to New York City to try to raise funds for his experiments, but he was so poor he didn't have the ferryboat fare. He would leave his umbrella with the ferryboat owner for security. While he was away, his family fished in a nearby lake to keep from starving.

Although Goodyear hadn't yet found the secret of perfect rubber, he produced many products that were better than those being manufactured by others. He made piano and table covers, maps of sheet rubber, aprons, curtains, beds, and even tents. The *Boston Courier* printed a few copies of their newspaper on sheets of Goodyear's India rubber and displayed them at their offices for everyone to see. Goodyear was even able to open a small salesroom to sell the products he made.

Goodyear really believed in his products. He would talk about rubber to anyone who would listen. He even took his patent applications himself to Washington, D.C. and gave samples to President Jackson and other important politicians. To advertise his products he wore rubber clothes when he went out. The skirts of his rubber coat were covered with dark spots where he had touched the coat to the stove to show everyone how tough it was. A local townsman was once asked how to locate the eccentric inventor. The man replied, "If you meet a man who has on an India rubber cap, stock, coat, vest, and shoes, with an India rubber money purse without a cent of money in it, that is he."

Eventually Goodyear went into partnership with a factory owner who had been experimenting with sulphur as a means of improving rubber. Goodyear became convinced that sulphur was the answer; he just didn't know what to do with it. Again, the rubber was improved but still not perfect. He found this out when he got a government contract for 150 mailbags. They turned out beautifully. They were smooth and waterproof and they looked good. The final test was their withstanding weather changes. Goodyear hung them by their handles. When he came back later, some were sagging so badly they had reached the floor. He was a failure once more.

One day he was holding a piece of rubber when it accidentally brushed against a hot stove. It suddenly looked like leather. Goodyear couldn't figure it out. Heretofore his rubber had always melted when it was exposed to heat. Then it dawned on him that very high heat would have to be used. Everyone had been stopping the experiments before the right temperature had been reached.

He heated another piece of rubber in front of an open fire with the same results. He nailed the piece outside the kitchen door in freezing weather. In the morning he expected to find it hard as a rock. Instead it was as flexible as when he had put it out. He had discovered the secret of making perfect rubber, a process that became known as

Sketch of Charles Goodyear.
(Library of Congress)

vulcanization, named after the Roman god of fire.

Unfortunately, Goodyear had gone around town boasting about his rubber for so long that few people paid any attention to his new discovery. They remembered the closed factories and the mailbag disaster. His own brothers refused to believe that his latest discovery was important.

By then Clarissa and the children were destitute. Charles sold his children's schoolbooks to buy food and they had to search nearby fields for fuel and half-grown potatoes. When his two-year-old son, William, died, they could not afford a proper burial, so they took him to the graveyard in a wagon. A neighbor once said, "He was so poor he couldn't be trusted to pay back an ounce of tea."

Life for the Goodyear family became so desperate that Charles finally set out through a freezing snowstorm to the house of Oliver B. Coolidge, president of the closed Eagle Rubber Company. Mr. Coolidge took pity on him and, in exchange for a thousand pounds of rubber, provided the family with funds to keep going. Mr. Coolidge did not think he would ever get his money back. He said he did not consider the rubber very good security.

Goodyear, on the other hand, later said he was never ashamed of begging from friends, neighbors, or even strangers. He said he felt it was his duty "to beg in earnest, if need be, sooner than that the discovery should be lost to the world and to himself."

Eventually, through begging, he got a factory in which to carry out his experiments. The owner of a dye factory said Goodyear could use the company machines. Later, however, the owner complained that Goodyear's experiments left the machines dirty and that the workers laughed at the experiments. One worker later said, "He used to be fussing about with those samples and I thought it was silly and like boy's play, and I used to laugh at it."

When Goodyear was asked to leave the factory, he begged the owner of a patent leather company and the village blacksmith to let him use their ovens. When these ovens were busy, he worked over open fires made from brushwood in the fields.

Finally Goodyear's efforts paid off. He found businessmen who believed in his process and he was able to continue his experiments and provide for his family. Each week of concentrated work brought more improvements to his rubber. On June 15, 1844, five years after he had discovered vulcanization, he finally received a patent for his process.

After Goodyear's patent was published, many others wanted to take credit for its discovery. People at the general store in the town where he lived said it was their stove that the rubber had accidentally touched when he discovered vulcanization. The local department store said it was theirs. Inventors from all over the United States, and eventually the world, declared they had invented vulcanization long before Goodyear had.

Goodyear tried to ignore these claims, but soon he was losing so much money from people stealing his idea that he was forced to go to court. The Shoe Association filed a lawsuit in Goodyear's name and the result was called the "greatest American business lawsuit of the nineteenth century." Newspapers called it "The Great India Rubber Case."

The Shoe Association hired a famous lawyer who was at that time secretary of state. Daniel Webster was offered ten thousand dollars to take the case, and five thousand more if he won. On September 28, 1851, the judge declared that Goodyear was the true inventor of vulcanization.

Meanwhile Goodyear had not been idle. In May of that same year he spent thirty thousand dollars to build Goodyear's Vulcanite Court for the

Great Exhibition in London. It was important that he made a good showing at the exhibition because the British had given the patent for vulcanization to someone else. He wanted to show the British what they had missed. His exhibit took up more space than all the other American displays put together. It consisted of three rooms, and everything in them was rubber—the walls, roofs, furniture, carpets, and drapes. Over six million people filed through to see "Goodyear's Vulcanite Court" and the inventor became famous overseas as well as in the United States.

After that it seemed as if ideas would not stop popping into Goodyear's head. By 1853 he had thought of nearly a thousand uses for rubber. Some of the things he invented—windmill sails, waterwheels, a rubber sail for ships, and the whale spring (a spring used to ease the strains on the cables that held whales during rough weather)—are no longer used. But many of the products we think of as modern were actually suggested by Goodyear 132 years ago. Rubber bands, machine belts, hoses, boats, footballs, life preservers, and overshoes are just a few examples.

Some of Goodyear's inventions were made before the inventor even knew what they would be used for. For instance, he invented the hot water bottle, but he thought it would be some sort of musical instrument. He thought of the water bed but believed only the lame and sick would use it. He fastened a rubber sponge on a wood block with a long handle and called it "squealgee," which was a nautical term, because he thought it would be used for scrubbing decks on ships. Today it is used for things like cleaning windows.

The people who worked with Goodyear never knew what he would come up with next. His whole world was rubber. His watch was encased in a rubber watchcase. He used knives with rubber handles. A rubber picture frame surrounded his portrait, also painted on rubber. His office doorplate was made of rubber and when he wrote a book on the many uses of rubber, a few copies of it were bound in rubber.

Not all of Goodyear's inventions became popular. He produced rubber maps and globes that could be used as tablecloths, bedspreads, or rugs so that children could be educated when they were eating, sleeping, or lying around the house. He made up a mattress cover that could be tied up tight and used as a life raft. Air-cooled boxing gloves never became popular, nor did the imitation buffalo robes he tried to sell. He designed a "fording dress" so people could keep dry when crossing a river on being baptized. He designed water bags which he said could be hung under the eaves to catch rainwater from the gutters (houses did not have running water in those days). The rainwater would travel through rubber tubes to all the rooms in the house. This never caught on.

It has been said that Goodyear foresaw almost every rubber product we use today. Only those used on such future things as airplanes and automobiles, which hadn't yet been invented, were safe from Goodyear's busy brain.

Goodyear became world-famous, but he never forgot the terrible times he and his family had gone through. He remained eccentric and unconventional. While he was in London it was not uncommon for him to drive his coach with his coachman riding inside. He remained thrifty the rest of his life, remembering the time he had spent in jail for debts, but he was always generous to his wife and children. His children said he often came home so loaded with presents he must have bought out entire stores.

His wife, Clarissa, who for thirty years had stood by him when no one else would, died in March 1853. The lonely Goodyear married again the next year. He had three more children, but only one survived him.

Goodyear, who had never been very healthy, lived a quiet life in Washington, D.C., until 1860. He learned that his daughter Cynthia was gravely ill in New Haven, Connecticut, and tried to get to her even though he was very ill himself. Accompanied by his physician and some family members, he arrived in New York and was told his daughter had died the previous day. He was too ill to make the rest of the trip to New Haven so he checked into a hotel. A month later he died.

3

Other Early
Inventors

CHARLES GOODYEAR died in 1860, but if he
had lived to see automobiles and airplanes, he
would have seen his invention become more important than ever before. He would have seen the
early wagons, carriages, and automobiles riding
smoothly on rubber wheels instead of on steel-
covered wooden wheels. He would have seen these
vehicles traveling smoothly over asphalt and rubber pavements instead of on dirt or rough cobble-
stones.

In the 1840s an Englishman named Robert
Thomson had attached an inflated canvas and
rubber tube to a carriage wheel to make the carriage ride more smoothly. He patented it as an
"elastic belt." In 1887 a horse doctor in Belfast,
Ireland, improved on Thomson's invention by tak-
ing a long, slim tube of thin rubber and cementing
the ends together. He attached a nipple from a

baby bottle and inflated the rubber with air, using his son's football pump. He put the tube on his son's bicycle wheel and found it worked perfectly. It allowed his son to ride on air instead of on steel rims.

Eventually these rubber wheels became known as tires because the wheel was *attired* in rubber. After the bicycle craze hit England and the United States in the 1880s and tires became even more popular, on Thanksgiving Day in 1895, a Chicago newspaper offered a prize of two thousand dollars to the winner of a fifty-five-mile race among six of the new "motor wagons." Only two finished in this first auto race. One of the drivers, a young, bearded inventor named J. Frank Duryea, had built the first successful gasoline engine motor vehicle in 1893. The car he used for the race was different from the first two vehicles he had built because it wore the first set of automobile tires in America. Duryea said it would enable man to ride on air.

Duryea's tires caused a stir among the spectators. In fact, right before the start of the race, a skeptical bystander drove a knife through one of the rear tires "to see if it was really solid or not." Frantic, Duryea bored out the cut with a hot poker and inserted a plug, using a couple of tablespoons of rubber cement to bond it. The tire was reinflated and Duryea went on to win the contest.

The first tires produced for bicycles and automobiles were not the sturdy, long-lasting rubber tires we know today. They were made of a fabric that looked like canvas coated with rubber. They were as smooth as a bowling ball, causing a car or bicycle to skid easily. Vehicle owners were lucky if their tires lasted a hundred miles without blowing out, and the speed they went was fifteen miles an hour at the most.

The first improvement on tires was the addition of treads. Charles J. Barlay in 1897 put a series of rubber buttons along the outer edges of a tire to give it more stopping power and make it skid-resistant. He called his invention "the won't skid tire."

Other improvements followed. Manufacturers began using carbon black in tires about 1918. Before that, they were white or red. With the addition of carbon black, tires were easier to keep clean. But more important, it made them tougher and longer-lasting. The invention of the balloon tire in 1923 meant that tires could be filled with more air. And the invention of the tubeless tire produced safer tires because they were less subject to blowouts. In the 1940s the Michelin steel-belted radial tire was developed in France and became the toughest tire ever. (A radial-ply tire is one in which the rubber strips of the carcass, the inner layer of the tire, are stretched from the inner to the outer edge.)

Other products made of rubber made it a very popular item, thanks to one man, Dr. Benjamin Franklin Goodrich. In 1870 B. F. Goodrich started the first rubber company west of the Alleghenies in Akron, Ohio. The company's first product was a cotton-covered rubber fire hose. Goodrich had watched a friend's house burn to the ground because the leather fire hoses used at that time had frozen and burst. He vowed he would make a fire hose that could withstand pressure and cold temperatures.

After the success of the new fire hose, the Goodrich company developed other important products. Goodrich was granted a patent on the first prestretched rubber belt. He developed new and better bicycle tires and eventually became known for developments in automobile tires.

By the 1890s it became the practice for anyone in the area having a manufacturing problem with rubber to take it to the B. F. Goodrich factory. For instance, Coburn Haskell was a good friend of B. G. Work, who was employed at Goodrich. One evening in the late 1890s the two men were discussing golf balls. Work declared he could make a better golf ball, so the two went to the factory the next day and together invented a new kind of ball made of rubber thread wound under tension and placed inside a gutta-percha cover. Today's golf balls are made by the same principles they developed and have revolutionized the game. Goodrich had some two hundred patents on rubber before his death in 1888.

All this resulted in a bonanza for the rubber industry.

4

India Rubber Fever

AFTER GOODYEAR'S DISCOVERY, "India rubber fever" hit the industrial world and rubber became so valuable it was known as "black gold." In fact, the quest for the new substance was considered the biggest boom since the gold rush in the Alaska Klondike. The great steel tycoon Andrew Carnegie once said, "I ought to have chosen rubber."

For a while the only place where rubber was cultivated was the Amazon Valley in South America where there were three hundred million virgin forest trees waiting to be tapped. Workers there hacked their way through jungles with machetes to reach them. It was so dark among the huge trees that at times they had to wear kerosene lamps on their heads as miners used to do. The workers slashed the trunks of the rubber trees with deep gashes and gathered the milky substance in gourds. When they ran out of containers, they smeared the thick sap on themselves and stripped it off when they got back to their huts.

The rubber was put into smoke huts where smudge fires kept it in a liquid state. Workers dipped long, broad-bladed wooden paddles into the rubber until it thickened even more. Then they wound it, layer upon layer, in a ball weighing as much as two hundred pounds. (Stones were sometimes added by unscrupulous traders to increase the weight so they could get more money.) The rubber was then taken to trading posts along the river.

By 1857 planters had brought in paddle-wheeled steamers, permitting trade with other countries—Bolivia, Ecuador, and Colombia—thousands of miles upriver. Rubber camps sprang up all across South America, bringing in thousands of workers. Greedy rubber traders sold shoddy goods—fake jewelry, cheap musical instruments, sewing machines, and guns that fell apart a few weeks later—to the natives and workers. Then they forced the natives to work off their debts. The natives went further into debt buying food and necessities. Some of them spent their entire lives working off these debts.

With profits the traders got from rubber, they built huge houses and city buildings. In Manaus, Brazil, they built streets of rubber so the carriages would not make noise. They built schools, museums, a library, a zoo, an electric plant, a sewage system, and an electric streetcar system. They built restaurants, cafés, bars, and gambling houses. Natives were hired to carry baby grand pianos to the jungle on their backs, but no one knew how to play the pianos. Caviar and champagne were brought in, as well as marble columns, gilded carvings, and alabaster for the buildings.

The Spanish and Portuguese controlled the lands where rubber was found and they were very strict about any rubber seeds being taken out of the country. In 1876, however, an English planter, Henry Wickham, succeeded in sneaking back to England about seventy thousand seeds from the rubber trees in Brazil. He planted the strange brown seeds in greenhouses at the Royal Botanical Gardens outside London. Only twenty-five hundred of the seeds survived, but these were enough to start plantations in other parts of the world such as Ceylon and Malaysia.

The eventual competition proved too stiff for many of the planters along the Amazon. With so many plantations elsewhere producing rubber, prices fell drastically. Many rubber towns, including Manaus, became ghost towns. Buildings and houses were abandoned. The docks were empty. The jungle took over again and jaguars and boa constrictors roamed freely through empty houses still containing elegant chandeliers and baby grand pianos.

With the advent of the bicycle and the automobile, rubber was needed more than ever, and the existing plantations were barely able to keep up with the demand. So some of the most important people in the United States became interested in finding new and different ways to produce rubber.

Thomas Edison, while in his eighties, called rubber the most complicated problem he had ever tackled. He believed the United States had only about a year's supply of rubber for the thousands of products manufactured annually. He did not think the United States should rely on foreign suppliers, so he began searching for a rubber plant that could be grown in Florida and other North American states with warm climates.

Edison built a ten-acre experimental garden in Florida and hired five traveling botanists and plant hunters to scour the Gulf Coast states for rubber-producing plants. They traveled in a camp-car rover that was equipped like a "miniature botanical laboratory" and roamed the highways and back roads of America searching for strange, unidentified plants. Other botanists were sent to Europe, Asia, and Africa.

Edison eventually had four thousand different varieties of tropical and semitropical plants in his Florida garden. Many of the plants were really weeds, but he found that even weeds could produce rubber. He worked closely with Henry Ford, the automobile maker, and with Harvey Firestone, owner of one of the largest rubber factories in the world. Ford, who owned and operated a

Thomas A. Edison and Harvey S. Firestone, one of America's leading tire manufacturers, discuss new rubber sources. *(The Firestone Tire & Rubber Company)*

thirty-two-hundred-acre rubber plantation near Savannah, Georgia, offered Edison land for field tests, the services of his experts, and shops for turning crops into rubber.

One of the plants Edison tried was the guayule plant. This shrub, rarely reaching higher than two feet, grew in a semidesert region of central Mexico. Instead of a milky sap, its rubber was encased in special cells. A process to extract its rubber was invented by an Italian chemist in 1900. Factories were built to grind up the plant and extract the rubber, but the process was long and expensive and produced very little rubber. Although Edison was successful in obtaining rubber from the guayule plant, not enough came from it to provide even a portion of the large amount needed by the growing automobile industry.

Although the eighty-year-old Edison worked fourteen hours a day, he failed to find an efficient and economical way to extract rubber from his plants. He had vowed to keep trying "at least until I have found all there is to know about rubber," but he died before he could find the solution to growing rubber in America.

Henry Ford agreed with Edison that America did not have enough rubber of its own. In fact, he said that the United States "didn't produce enough natural rubber to supply elastic for our sock supporters." Like Edison's American plantation, Ford's plantation in Florida did not produce enough rubber. He came to believe the solution was to increase the number of American-owned plantations abroad.

Experiments using guayule to produce rubber were tried over many years. *(Library of Congress)*

By 1910 tires had been improved considerably, as evidenced by this Rolls-Royce roadster. *(Harrah's)*

Ford built his own plantation in Brazil and sent a fleet of ships loaded with steel, cement, locomotives, medicines, foods, "everything from safety pins to sawmills," up the Amazon. The ships dropped anchor among the green jungles of the lower Tapajoz, an area inhabited by monkeys, parrots, and native Indians that was called "the last home of mystery" because archaeologists and scientists searched there for lost civilizations.

Ford built an entire city on fifteen thousand acres of land in the jungle. It included a refrigeration plant for storing meat and provisions for six months for two thousand people. It included a hospital and small houses with electricity and running water. He built schools, clubs, a tennis court, a golf course, a swimming pool and, of course, an experimental laboratory.

Ford hired Brazilian workers to plant a million and a half rubber trees. Unfortunately, American overseers tried to get the natives to work by time schedules. The workers were used to going to work at sunrise and stopping at midmorning when it became hot. They would resume work in the late afternoon when it was cool. The Americans antagonized the workers by feeding them strange foods such as cornflakes, spinach, and lettuce when they were used to rice and fish.

The first night the workers ate in the plantation cafeteria, they threw pots at the overseers and managers. The Americans had to escape to the company ships anchored in midstream while the natives went on an all-night rampage. The Americans sent frantic messages for help, but by the time a hydroplane landed with state troops aboard, the natives had settled down. There would be no more spinach and no more time clocks.

Meanwhile, researchers at the Firestone and Goodyear companies found that there was a sort of "rubber belt" around the earth's circumference. The belt extended about fourteen hundred miles south of the equator and seven hundred miles north of it. Almost all the rubber trees in the world grew within these latitudes.

Soon new plantations were built in the Malay Peninsula, Burma, India, Ceylon, Borneo, China, New Guinea, Dutch East Indies, Africa, and the Philippines. Thousands of natives were hired to tap the rubber trees and care for them. Thousands

more worked around the clock in factories processing the rubber. They soon found that just as with
cotton plantations in the United States, there
could be good plantations and bad plantations.
Some—like Firestone Park and Goodyear
Heights—included houses, insurance, doctors and
nurses, banks, schools, and recreational activities.

Other plantations were not so nice. They offered houses, easy hours, and fifty cents a day to
workers. But when the workers arrived, they
found they already owed the company for their
train fares, straw mats for sleeping, pans to eat
from, and cotton work outfits. If they tried to run
off, they were captured and everyone was called in
from the fields to see the escapees flogged. In these
bad plantations eighty workers lived in each of
the crowded *galeras* (prison huts) surrounded by
barbed wire in a cramped area sixty feet long by
thirty feet wide.

Edison once wrote, "Henry Ford, Harvey Firestone, and I were considering what this country
would do in case of a war which cut off our rubber
supply. Don't make any mistake about that war; it
will come. We may run along for a good many
years without it, but sooner or later the nations of
Europe will combine against the United States.
The first thing they will do will be to cut off our
rubber supply."

Edison's prediction came true in 1941 when
America entered World War II. The Germans, who
thought they had been thoroughly prepared for

the war, had forgotten about rubber. Early in the war they realized their mistake and tried to buy up as much as possible. They tried to smuggle rubber out of South America by plane or "runner" ships, and they became so desperate they tried to use steel tires with circles of small coil springs attached to the wheels of their vehicles instead of rubber tires. None of their substitutions worked.

Other nations were also caught unaware. Great Britain, which had supplied many other countries with rubber, had no stockpile whatsoever. Allied ships had to travel twelve thousand miles and more over ocean routes threatened by enemy ships to bring back rubber. Many ships were sunk or captured. The Japanese captured the main sources of natural rubber in Asia. Suddenly countries like Britain, Russia, China, Australia, Africa, and Canada were all relying on the United States for rubber. Like the Germans, the Americans realized that the country that had rubber ruled the war.

When the United States entered the war, all rubber immediately went to the war effort. Of the thirty-four thousand rubber-built articles under patent in 1941, twenty-four thousand went for the army, navy, and air corps—things like rubber mounts for antiaircraft guns, rubber cushions and mattresses for bunks, sponge rubber for the insides of tanks, huge tank treads, rubber oxygen masks for pilots, and a thousand other military products.

Movie star Judy Canova urges Americans to donate their rubber products to the war effort. *(Library of Congress)*

As Edison, Ford, and Firestone had predicted, there was not enough rubber to last six months. Americans were urged to donate all their rubber products to the war effort and they responded immediately. Federal agencies donated their spittoon mats. The Boy Scouts manned service stations, begging motorists to give up the rubber mats from their cars. A Los Angeles rubber company turned over five thousand tons of old tires, and President Roosevelt even donated his Scottie's rubber bone. This search for rubber was dubbed "the greatest scavenger hunt in history."

Frantically, American businessmen reopened plantations that had fallen into ruin in Brazil.

Manaus again became an active city. The largest fleet of riverboats in the world arrived to take out all the rubber possible.

Still, this was not nearly enough. The only answer was to develop a substitute or synthetic rubber. The four largest rubber companies were asked by the government to build synthetic rubber plants that would produce twenty-five hundred tons of synthetic rubber a year.

Until the war there had been little interest in synthetic rubber because natural rubber was so cheap. The first synthetic had been developed as early as 1860 when an Englishman, Grenville Williams, using heat, changed rubber from a solid into a liquid. He called his discovery "isoprene." In 1887 chemists discovered how to turn isoprene back into a rubberlike substance, and in 1909 a German scientist, F. Hoffman, took out the world's first patent for a synthetic rubber.

Researchers began experimenting with these older synthetic rubbers and with newer ones—neoprene, Buna, and Ameripol. Ameripol was discovered by Dr. Waldo Semon. He had been working for Goodrich for only a few months in 1926 when he discovered Koroseal, a synthetic product made of coke and limestone. Koroseal could be sprayed on cotton fabric to make waterproof material for raincoats, shower curtains, and baby pants. It was also fire-resistant, which made it an ideal insulation for electric wires in battleships and planes. Semon then turned his attention to

High school students from Washington Irving Vocational School for Girls bring in rubber products for the war effort. *(Library of Congress)*

discovering synthetic rubber. He eventually made eight thousand different rubbers, but the best was Ameripol.

Neoprene was discovered in 1931 by Reverend Julius Arthur Nieuwland, a chemist at the University of Notre Dame, and Buna was originally made by German chemists. By 1944 all four plants originally contacted by the president were producing eight hundred and fifty thousand tons of synthetic rubber to help win the war.

5

Rubber's Growth and Development

AFTER THE WAR U.S. government officials vowed they would never again rely on foreign rubber. They knew if they didn't develop their own supplies, they could be charged any price demanded by other countries. Rubber producers could even refuse to sell rubber to the United States, and the American economy, which relied so heavily on rubber, could be ruined.

Chemists and researchers continued to work on improving natural and synthetic rubber. Some of them became known as "stretch detectives" because they dedicated their lives to finding new uses for their products. For instance, when seven giant one-hundred-year-old trees on Thomas Jefferson's estate began dying, the "stretch detec-

Synthetic rubber looks like popcorn as it comes from the drying machine. *(The Firestone Tire & Rubber Company)*

tives" were called in. They made up a rubber product that could be poured into the holes in the trees so they would be saved. They produced a hose that could stand mud pressure of sixteen hundred pounds for oil drillers. And when the owner of a poultry farm needed a mechanical brooder to mother his motherless chicks, researchers made him a rubber hen.

Today there are thousands of synthetic rubbers on the market, but natural rubber is still grown as well. Most natural rubber comes from plantations in Malaysia, Indonesia, Brazil, India, Vietnam, and Africa. Workers there still go from tree to tree tapping the white sticky liquid.

Most of today's natural rubber comes from the Hevea tree which has shiny, oval dark green leaves. The tree also produces small, pale yellow blossoms. When the flowers drop off, the seedpods appear. When the pods are ripe, they explode with a loud bang. Seeds are shot some hundred feet away, where they root and become new trees.

The white sticky liquid is found in the tree's bark. It looks like milk and is, in fact, called latex from the Latin word *lac,* meaning milk. After the tree has been growing for six years, this "milk" can be tapped. Some of the trees continue to pro-

Rubber comes from a white sticky liquid found in the bark of the Hevea tree. *(The Firestone Tire & Rubber Company)*

Workers make fresh cuts in the bark of the rubber tree to get to the latex. *(The Firestone Tire & Rubber Company)*

duce latex for thirty years and full-grown trees can produce four to fifteen pounds of rubber a year.

Workers tap and collect the latex each morning before the sun comes up. They carry gouges with them and use the gouges to cut a narrow groove in the tree about four feet above the ground. A metal spout is put at the lower end so the latex can flow out. They tie small cups to the tree below the spout to collect the latex.

Latex is taken from the field trucks to the processing plant by truck. *(The Firestone Tire & Rubber Company)*

The latex is shipped in one of three ways: (1) in liquid concentrate form, (2) as solid, smoked rubber sheets, or (3) as crepe rubber. Smoked rubber sheets are made of dry rubber. Water is removed from the latex and the latex is mixed with certain acids. This process causes it to solidify into spongy water-soaked slabs (or ribbons)

To make dry rubber, coagulated rubber is washed, shredded, and dried in a machine called an extrusion dryer. *(The Firestone Tire & Rubber Company)*

which are passed through rollers. More water is squeezed out and the result is a slab which is smoked and dried in huge houses.

Crepe rubber is similar to smoked rubber, but the crepe rubber is allowed to dry in the air instead of in smokehouses. It is used to make things such as the soles of canvas shoes.

The rubber goes through three steps once it arrives at the rubber factories. It has to be mixed with other ingredients. Then it has to be given a shape. It has to be vulcanized and, before the ingredients are mixed, the natural rubber has to be masticated, or chewed, to make the material softer and more malleable. Then all the ingredi-

Some sheets of rubber are hung in a hot room to dry. *(The Firestone Tire & Rubber Company)*

To make concentrated latex, separators are used to extract the water. *(The Firestone Tire & Rubber Company)*

After the final mixing, the rubber is rolled into a continuous sheet ready for further processing. *(The Firestone Tire & Rubber Company)*

ents are kneaded together, usually by sending them through rollers.

The second step, giving the rubber a shape, is done in a number of ways. One way is called extrusion. The rubber is forced like a sausage through shaped dies. The result can be flat strips or tubes, hose linings, cable coverings, or any number of similar products. If the manufacturer wants flat sheets, the rubber is sent through rollers.

In the mixing operation, rubber, carbon black, sulphur, and
other materials are carefully measured into a large machine.
(The Firestone Tire & Rubber Company)

Crepe rubber is pressed into bales and loaded on a ship. *(The Firestone Tire & Rubber Company)*

The third step, vulcanization, can be done in a sort of pressure cooker or with induction heating or heated rollers. It can also be done by molding. With molding, the rubber is put into a metal mold and vulcanized while it is inside.

Manufacturers today use these methods to make almost forty thousand different products. Some are used by the medical industry. Surgeons' gloves, hoses to supply blood and oxygen, and rubber aprons for X-ray technicians and patients are all made of rubber. Rubber is used for containers for blood plasma, antibiotics, and other medicines, as well as everyday pills such as vitamins. It is used for artificial heart valves and kidney machine parts.

Industries use thousands of tons of rubber each year. Rubber belts and hoses keep factories and mining machinery operating. Conveyor belts

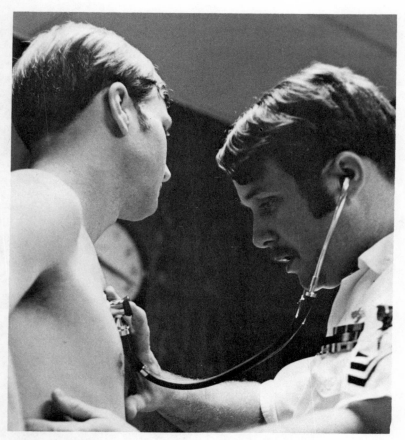

Rubber is used in many medical products. *(Official U.S. Navy photograph)*

move coal, minerals, ores, building materials, foods, and hundreds of other products to the markets. The food industry uses inflated rubber bags between plates that keep food cargo from shuffling around in railroad cars.

The building industry uses bridge expansion parts made of rubber. Roofing sheets are laid over rubber insulation and layers of rubber are laid in

Construction vehicles with their gigantic tires require large amounts of rubber. *(Official U.S. Navy photograph)*

44

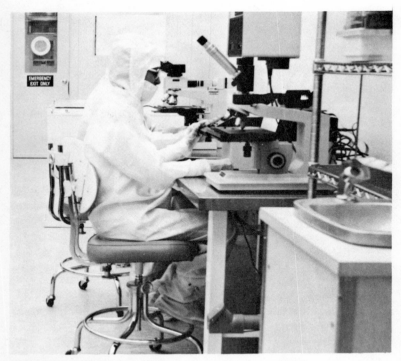

Motorola workers wearing rubber "bunny suits" use a micros-
copy lab to produce sterile products which make our lives
easier. *(Motorola Semi Conductor Products Sector)*

concrete as a barrier to moisture. Rubber fabric is
used to make temporary dams to prevent flooding.
Large acid tanks that use chemicals are lined with
rubber to prevent corrosion. Workers wear protective clothing made partly of synthetic rubber.

The farm industry uses rubber in milking
machines, plows, cultivators, seeders, combines,
hay balers, and cotton, vegetable, and fruit machines. Crops are protected from insects and dis-
eases by insecticides and fungicides made by the
rubber companies.

Rubber hoses are used to refuel fighter planes in midair. *(Official U.S. Navy photograph)*

Thousands of rubber products are found in every home today from rubber bands to rubber duckies. Raincoats, rain boots, and bathing caps keep our bodies dry. Some wallpapers are rubber treated, and latex paints have a rubber base. Mattresses, chairs, and couches may have foam rubber padding. Rubber insulation and parts are found in ovens, refrigerators, mixers, and blenders. Lawn mowers move on rubber wheels, and garden hoses carry water to thirsty plants. Our recreational time is filled with rubber products—beach balls, bowling balls, tennis balls, footballs, and golf balls. Some swimming pools are made of rubber.

A Navy underwater demolition team practices defensive techniques. *(Official U.S. Navy photograph)*

The people who protect our lives also rely on rubber. Firemen wear rubber outfits and use rubber hoses. Fire trucks, police cars, and ambulances roll to emergencies on rubber tires. The armed forces use rubber-based materials in protective shields for such things as gamma radiation. One fuel tank, called the pillow fuel tank (because it looks like a huge inflated pillow), can refuel up to nine jet fighters in twenty-seven minutes in the air. Sonar systems use rubber hoses, expansion joints, and driving belts. A rubber pulsating "skin" is put on the outside hulls of submarines and ship bottoms. Navy frogmen explore the seas and oceans in rubber outfits.

Synthetic rubber and fabric skirts on Hovercrafts control the air cushions that speed them on their way. *(Official U.S. Navy photograph)*

The biggest user of rubber today, however, is still the transportation industry. Manufacturers use rubber to build huge airships like the Goodyear blimp. Rubber weather stripping is needed for hangars to protect the planes inside. Some rubber is used for warplanes to make them self-sealing so that if a bullet strikes, the people inside are protected. These seals are made of several layers of thin rubber alternated with layers of nylon or other strong man-made materials. When a bullet strikes, the rubber swells up and seals closed.

Synthetic rubber and fabric skirts on Hovercrafts control the air cushions that speed them on their way. In England oil barges are made of woven nylon fabric coated with rubber. And synthetic rubber dock fenders are used in ports all over the world to protect the wharves from the huge ships that constantly bump up against them. Rubber cushions protect ship engines. And the edges of portholes and doors are lined with rubber to keep out the water and wind. Rubber is made into sealants for jet runways, bridges, dams, and reservoirs. Rubber added to asphalt makes our highways and roads safer and longer-lasting.

Automobiles, taxis, buses, and trucks have over six hundred parts made of rubber. Rubber is used in transmission belts, windshield wipers, floor mats, parts of the motor, insulated wires, and weather stripping around the doors, windows, and trunks. Hydraulic safety brakes, power steering, radiators, and air-conditioning all need rubber hoses to work. Rubber foam is used in upholstery and as safety padding.

Huge rubber dock fenders keep boats from ruining wharves when they scrape up against them. *(Official U.S. Navy photograph)*

49

Airplanes must be completely sealed so that no wind is allowed in and no air from inside is allowed to escape. Rubber provides that protection. Rubber tires are made for jets that can withstand thousands of pounds of pressure, friction, and impact upon landing. Rubber mats are put on the floors, and foam rubber is used for seat cushions. The baggage compartments have rubber panels to protect luggage (which is sometimes made of rubber), and rubber-coated insulation wires and cables keep the engines going. Deicers are made of rubber-covered inflation tubes. They are attached to the outside where ice might collect during flight. The tubes inflate and deflate, causing the ice to crack. It is then carried off in the airstream.

The space age has brought about new uses for rubber. Rubber and fabric hoses are needed to fuel rockets. Parachutes reinforced with rubber bring astronauts safely to earth during reentry. Metal rocket cases are insulated with rubber to protect missiles from the heat of burning fuel. Even part of the solid fuel burned in booster rockets contains rubber. Rubber-based materials form a protective shield for rockets and space vehicles against neutrons in space.

Rubberized fabrics have been used to build space stations, and synthetic rubbers resistant to heat have been used for supersonic aircraft and space satellites as well as for manned capsules and missiles. A missile insulation liner has been

A Royal Australian Navyman, a sealab III man-in-the-sea aquanaut, tests a rubber diving suit. *(Official U.S. Navy photograph)*

called the "world's largest complete rubber product." It is two hundred and sixty inches in diameter and weighs ten thousand pounds. Some of the huge space platforms contain rubber and the first tires on the moon were made of rubber strong enough to withstand the rough moon terrain.

Rubber pressure suits for astronauts were developed by a flyer named Wiley Post in 1934.

Unfortunately they were very stiff, making it difficult for the wearers to move their arms and legs. In 1952 an engineer named Russ Colley was watching a tomato worm in his garden and got the idea of making space suits with joints that allow more flexibility. Without the Mark IV suit that developed from Colley's idea, pilots and astronauts would black out and die from the intense air pressures. Today's space suits are made from shiny, flexible nylon and rubber and are coated with aluminum.

Tires still make up the biggest percentage of rubber used today. More than half of all the rubber used today, including both natural and synthetic, goes for tires. These range from the tiny tires found on model cars (more than 105 million of these tires were made for 27 million battery-powered vehicles in 1984) to gigantic mining dumpster tires that may reach eleven and a half feet in diameter and weigh three tons each. Some companies spend millions of dollars a year on tires so huge they have to be lifted by cranes to be changed. Giant earth-moving vehicles travel over sharp rocks, stumps, and ledges on enormous rubber tires. And tires built to withstand temperatures ranging from below zero to over a hundred degrees Fahrenheit travel to places like the Arctic Circle or the African deserts.

What will the future hold for rubber? Rubber will continue to exert a major influence in the

A synthetic rubber plant works overtime to improve the quality of our lives. *(The Firestone Tire & Rubber Company)*

world economy, and the demand for it is rising as newer nations become more industrialized.

More important, rubber will continue to contribute to the quality of life on an already crowded and noisy planet. We take for granted our smooth-running automobiles, our quieter-than-cobblestone streets, our silent appliances, and luxurious padded carpets. But each year better developments in noise control result in quieter bridges, floor coverings, engine mountings, and antivibrator units using rubber. Adding rubber cushioning to railway wheels and tracks, especially in cities, will make them a little quieter and will add safety factors as impact-resistant bumpers, shock-absorbing interiors, and crash-resistant fuel tanks are developed. Recent experiments prove that the addition of as little as 3 percent of rubber to asphalt can produce road surfaces that are longer-lasting, safer, and much quieter. Goodyear's crusade to make our lives safer and more peaceful will continue to be accomplished.

Index

Academy of Science, Paris,
  France, 2
Africa, 27, 34
Airplanes. *See* Rubber, use of
  in airplanes
Akron, Ohio, 19
Amazon River, 26
Amazon Valley, rubber trade
  in, 20, 21–22
Ameripol, 31, 32
Arctic Circle, 52
Automobiles. *See* Rubber, use
  of in automobiles

Barclay, Charles J., 18
Belfast, Ireland, 16
B.F. Goodrich Co., 19, 31
Bicycles. *See* Rubber, advent of
  bicycles and
Borneo, 27
Boston, Mass., 3
Boston Courier, 8
Boy Scouts, 30
Brazil, 21, 22, 26, 27, 30–31, 34
Buna, 31, 32
Burma, 27

Calendar (rubber processing), 4
Caoutchouc, 2
Carbon black, 18
Carnegie, Andrew, 20
Ceylon, 22, 27
Chaffee, Edwin M., 3–4
China, 27
Colley, Russ, 52
Columbus, Christopher, 1, 2
Condamine, Charles Marie de
  la, 2
Coolidge, Oliver B., 10–11
Cortez, Hernando, 1, 2
Crepe rubber, 38

Duryea, J. Frank, 17
Dutch East Indies, 27

Eagle Rubber Co., 11

Edison, Thomas, 23–25, 28, 30
Elastic belt, 16
Ether, 3
Extrusion, 40

Faraday, Michael, 5
Firestone, Harvey, 23, 28, 30
Firestone Co., 27
Firestone Park, 28
Ford, Henry, 23–24, 25–27, 28,
  30
French Guinea, 2
Fresneau, Francois, 2

Galeras, 28
Germans, rubber supply and,
  28–29
Golf balls, improvement of, 19
Goodrich, Dr. Benjamin Frank-
  lin, 18–19
Goodyear, Charles, 4, 5–15, 20,
  54
  death of, 15, 16
  discovers vulcanization, 9–10
  other inventions of, 13–14
  poverty of, 7, 10, 11
  receives patent, 12
Goodyear, Clarissa, 6, 7, 10, 15
Goodyear, Cynthia, 7, 15
Goodyear, Ellen, 7
Goodyear, William, 10
Goodyear Co., 27
Goodyear Heights, 28
Goodyear's Patented Spring
  Steel Hay and Manure
  Fork, 5
Goodyear's Vulcanite Court,
  12–13
Great Britain, rubber shortage
  and, 29
Great Exhibition, London,
  England, 13
Guayule plant, 24–25

Hancock, Thomas, 3
Haskell, Coburn, 19

55